Know Your Goats

Jack Byard

Old Pond Publishing

First published 2014

Copyright © Jack Byard, 2014

The moral rights of the author in this work have been asserted.

ISBN 978-1-908397-88-1

A catalogue record for this book is available from the British Library

Published by

Old Pond Publishing
5m Enterprises Ltd
Benchmark House
8 Smithy Wood Drive
Sheffield
S35 1QN

www.oldpond.com

Book design by Liz Whatling
Printed and bound in China

Contents

Acknowledgements

I would like to acknowledge the help and advice that has been crucial in putting this book together.
Ólafur R Dýrmundsson, PhD. National Adviser on Organic Farming and Land Use. Iceland.
Antonietta Kelly of the Italian Trade Commission.
Tony Harman, Maple Leaf Images, Skipton, North Yorkshire.
Without their help and that of many dozens of farmers and breeders, this book would have foundered.
I do not accept responsibility for any mistakes; the responsibility for these rests firmly with my granddaughter Rebecca, who is now old enough to know better, and my young friends Sophie and Lauren.

Picture Credits

(1) Margaret Wolfs; (2) Beverly Currie; (3) Moritx Smaltz; (4) Geoffrey M. Trotter DTM; (5) Levens Hall & Gardens, Kendal, Cumbria, www.levenshall.co.uk; (6) Martin Doyle, Bilberry Goats, Waterford; (7) Paola Lorini - Natura Mediterraneo; (8) Adrianne Bell; (9) Melrose Dairy Goat Stud; (10) Richard Scrivener, Rare Breed Goats; (11) www.plamp.cz; (12) Martin Lehmann; (13) Roelf de Jonge; (14) Charles Rinehart; (15) Salvatore Pipia, The Italian Goat Consortium, www.goatit.eu; (16) Martin Byrne – GSPCA; (17) Halla Eygló Sveinsdóttir; (18) John Hancock; (19) Kinder Korner Goats; (20) Linda Østbye; (21) Lance Hays, Honey Doe Farm; (22) Istituto di Zootecnica Generale - Universita di Palermo; (23) White Fireweed Farm – Becky and Tim Hammond; (24) Ironwood Hill Farm; (25) Beate Milerski; (26) Tania Salreta; (27) Marshview Pygmy Goats; (28) Patricia Holmberg – Little Bit Acres Farm; (29) Franck Rimaud; (30) Brian Goodwin; (31) Jeannette Beranger, The Livestock Conservancy; (32) John Hancock; (33) Morgan Fredericks; (34) Abra K. Serretti; (35) Milan Korinek; (36) Willowbank Toggenburg; (37) Paul Asman and Jill Lenoble.

Foreword

Goats are a diverse range of beautiful (and sometimes rare) animals. Billy or nanny goats are the source of many a children's story - who hasn't read *The Three Billy Goats Gruff* to their children?

Goats were taken into the human fold over 10,000 years ago and most have been wandering the fields, hillsides and mountains since Adam were a lad. From some we obtain the fibre to create exquisite mohair and cashmere clothing. From others we obtain possibly more mundane (but equally important and extremely healthy) meat and milk; the latter being frequently turned into mouth-watering cheeses, bringing joy to me and a living to many farmers.

Goats are browsers and prefer unwanted brush, briar and weeds to grass, their lips and tongues choosing only the tastiest plants. Extremely intelligent and curious, they also are experts at escaping from 'secure' fields. Apart from keeping your fields looking good, they can make excellent pets.

Jack Byard 2014

Anatolian Black

Native to:
Turkey

Now found:
Mediterranean
and Aegean regions

Description

This has long, coarse flat hair - normally all black, occasionally brown, grey or pied - and large, long, droopy ears. Its weight is 45-198kg (99-198lbs), and its height is 76.2-101.6cm (30-40").

Goats like the Anatolian Black have been bred and developed for almost 3,000 years up to the present day. The Anatolian is believed to have originated in Turkey, and with its long hair and large droopy ears, is classed as the Syrian type of goat. These mountain breeds are normally found around the Mediterranean and Aegean regions and are well-adapted to surviving the wild weather and sparse feed. The long, thick, hairy overcoat insulates the Anatolian against the cold, and the breed also has a tremendous tolerance to disease. The breeders set up camp in the mountains with their herds, where the Anatolian browse on grass, shrubs and bushes - goats are brilliant mowing machines. These calm, gentle goats are bred for their fibre, milk and meat.

Angora

Native to:
Himalayas

Now found:
Worldwide

Description

The colours are red, shading to tan, brown, black, grey and white. The underbody and inside of the legs are covered with curly or wavy mohair. The ears are droopy, and both billies and nannies have gently curving horns. Its average weight is 40kg (90lbs).

Goat hair or fibre has been used for clothing for over 3,000 years. Mohair is the silky, lustrous and hard-wearing fibre from the Angora goat. Originally coming from the Himalayas, the goats were herded to Ankara by Suleiman Shah, who was fleeing from the legendary Genghis Khan. Angora is a derivation of Ankara and Mohair from the Arabic 'Mukhayua'. The finest Mohair comes from the six-month-old kids; the hair coarsens as the goat ages. Records show that the Angora reached these shores in the 1500s, but did not survive, and the Angora owned by Queen Victoria suffered the same fate. It was not until the 1980s that they were truly in the British Isles.

3.

Appenzell

Native to:
Appenzell, Switzerland

Now found:
Appenzell and Gallen,
Switzerland

Pure white with medium-length hair, this goat is usually hornless. It weighs 45-65kg (99-143lbs), and its height is 75-85cm (29.5-33.5").

Appenzell goats are found in the cantons of Appenzell and Gallen in Switzerland and are used mainly for their milk. This is an unbelievably beautiful pure white goat. In the1980s it was infected by a terrible disease, Caprine arthritis encephalitis (CAE), and was on the point of extinction. In 2000, only 677 were registered, and in 2011 this had increased to 1479. These increases have been due to the help and support of the FOA, the Swiss Goat Breeders Association and ProSpecieRara, the Swiss Preservation Foundation, as well as cheesemaker Matthias Koch. Matthias created a new cheese made solely from Appenzell milk, which has been extremely successful. With a guaranteed market for the milk, the farmers are increasing their herds to keep up with demand. Traditionally, a herd of Appenzell lead cattle to and from the summer pastures. The beautiful Appenzell is recovering.

Arapawa

Native to:
Descendant of UK breed
introduced to NZ

Now found:
Worldwide, although very rare

Description

This has a short, fluffy coat with shaggy leggings. Colours include black, brown, tan, fawn and creamy white, and tri-coloured with black or dark brown badger stripes on the face. Billies have flattened, sweeping horns. Nannies have shorter, rounder horns which curve backwards. Its height is 63.5-76cm (25-30").

Small but beautifully formed, the Arapawa was introduced to New Zealand by Captain Cook, who landed the first two goats on 2nd of February 1773. In 1777, he presented a Maori chief with a further two. It is accepted that the Arapawa are the descendants of these four Old English Milch goats. In 1970 the New Zealand government decided to cull them, as they believed they were damaging ancient woodland. Betty Rowe spent a lifetime battling on behalf of the Arapawa. A number were taken off the island to safety, to breed elsewhere. There are less than 300 domesticated Arapawa worldwide, with approximately to 50 in the British Isles. The battle continues.

5.

Bagot

Native to:
Probably the Rhône Valley

Now found:
UK, although rare

Description

The Bagot has a black head and shoulders, and the rest of the body is white. The hair is long and shaggy. The long, twisting horns sweep backwards. Its height is 76.2cm (30").

Presented to Sir John Bagot in the 1380s by King Richard II, the Bagot is one of the oldest registered goat breeds in the British Isles, arriving on these shores courtesy of the returning Crusaders. These semi-feral goats have browsed the parklands of Blithfield Hall in Staffordshire for over 600 years. In WWII the herd were found guilty of damaging vital crops and sentenced, by the War Agricultural Executive, to be destroyed. Eventually, it was agreed that the herd would be reduced to 60; that number was retained for the remainder of the war. A number of black-and-white goats wander the hills of Wales; they are not Bagots, escapees from the Hall having a night out (need I say more?). Commercially, they have nothing to offer but their beauty.

Bilberry

Native to:
France

Now found:
Ireland

Description

With short legs but a large, strong body, the Bilberry has large, curling, wavy horns; a long, shaggy silky coat; a long beard; and a fringe that covers its eyes. Its average weight is 35-75kg (77-165lbs).

A very rare and gentle breed, this is very different to any other wild goat in Europe. In the year 2000, only seven remained. However, with the help of Martin Doyle, the organizer of the Bilberry Goat Heritage Trust and the Irish Wildlife Trust, there are now 89. Thought to have been brought to Waterford Quays in Ireland in the 17th century by Huguenots escaping religious persecution in Europe, these goats were put out to graze on common land at Bilberry Rocks. They have been moved along many times for many reasons, often because people did not want them there. They now have a permanent home where they can live undisturbed. This important piece of Irish heritage must not be allowed to fade into the sunset.

Bionda dell' Adamello

Native to:
Valle di Saviore, Lombardy

Now found:
Across Valcamonica region
of Italy

Description

This has long, fine, light brown hair and regular patches of white on the head and legs. The underbody and ears are white, with white stripes each side of muzzle. Its weight is 55-75kg (121-165lbs), and its average height is 74cm (29").

The Bionda dell' Adamello originated in Valle di Saviore in Lombardy, before spreading out into other local valleys. It takes its name from the Italian for blonde/fair. In the mid-20th century it was on the verge of extinction; in 1995 there were barely 100. However, with the help of the farmers and R.A.R.E. Association, this has risen to over 4000, and is continuing to increase. It would be tempting to intensively breed the Bionda to take advantage of the increasing popularity of the smoked Fatuli and Mascarpi whey cheese made from their milk, but Alpine breeds do not lend themselves to intensive breeding. In spring and summer they are taken up the mountains to graze, as nature intended.

Boer

Native to:
South Africa

Now found:
Worldwide

Description

Usually with short, white, smooth hair, a chestnut-brown head and floppy ears, the Boer has a short, stocky body with a broad chest. Its weight is 90-135kg (200-300lb).

This goat was originally developed in South Africa in the early 20th century. The Boer (Dutch for 'farmer') is believed to have been created using the goats from the Namaqua Bushmen and the Fooku tribe, and crossing and improving them with European and Indian breeds. It is the only goat breed in the world to be bred specifically for meat, and is entirely different in appearance to dairy breeds, having a stocky and solid appearance. The Boer was imported into the British Isles in the mid- to late-1980s, and despite being bred in a warmer climate, has adapted well to the vagaries of the British weather. It is now well-established, and has an excellent export market. It will quickly clear a pasture of weeds, improving it for grass-feeding stock.

British Alpine

Native to:
Britain

Now found:
Worldwide

Description

This goat is tall and graceful with a short, fine, glossy black coat and white or cream facial and leg markings. The ears are erect, pointing slightly forward. Its height is 83-95cm (32"-37") at the shoulder.

The two types of Alpine goat (Capra hircus - just showing off) are British and French. A Swiss goat with the grand name of Sedgmore Faith who lived in the Paris Zoo had the beautiful and distinctive black-and-white markings of her breed. In 1903 she was brought to England and crossed with a Toggenburg of similar colouring. All the kids had the beautiful Swiss markings. More breeding and refining took place; the British Alpine had arrived and was here to stay, and Sedgmore Faith was the grandmother of them all. An area to forage, a supply of hay, and a muesli-type supplement is happiness for the British Alpine, which produces a good supply of high quality, easily digestible alternative milk for all the family.

British Primitive

Native to:
Britain

Now found:
Britain

Description

This has long, thick, dense hair coloured mainly white, grey, and black. The large horns form a scimitar twist. Its height is 55.9-68.6cm (22-27"); its weight is 100-120kg (220.4-264.5lb).

'British Primitive' is the name that covers the breeds previously known as Old English, Scottish, Welsh, and Irish, together with British Landrace breeds - not forgetting the Old British Goat (I think I used to work for her). The BP is a descendant of the goats bred by farmers of the Neolithic, or New Stone Age. The Vikings, Saxons or Celts would have bred this hardy animal; a born survivor, protecting itself and its young against various predators, as well as having the ability to survive the harshest of weather on a poor and meagre diet. The BP provided the farmer and his family with milk, meat, skin and fibre for clothing, and tallow for heating and lighting. Nothing was wasted. They are now used for scrub clearance and conservation grazing.

11.

Brown Shorthair

Native to:
Czech Republic/Germany

Now found:
Czech Republic/Germany

Description

This breed has a short, glossy, brown fibre coat with a black spinal stripe starting with a triangle behind the ears and finishing at the base of the tail. The underbody, lower legs, hooves and inner of the upright ears are black. Its weight is 45-80kg (99-176lb); its height is 65-80cm (22.5-31.5").

This is a popular breed in the Czech Republic and Germany. It was developed in the late 19th and early 20th centuries on the border of the two countries using local brown goats and Brown Alpines. Great importance was placed on maintaining the hardy and adaptable traits and ability to survive in the worst of climatic conditions. In winter, when temperatures can drop to -20C (4F), they shelters in barns. In summer they typically browse at 800m, feeding on bushes and grass; a great favourite is raspberry canes. This is a high-yield commercial dairy goat, and its milk has many uses, including cheesemaking.

Chamois

Native to:
Europe

Now found:
Widespread

Description

The summer coat of this breed is reddish brown with a dark dorsal stripe; in winter, the coat is blackish brown. It has a brown face with a darker stripe running from the muzzle to the eyes. Both sexes have horns. Its weight is 54.5-68.18kg (120-150lbs); its height is 71-76cm (28-30").

The beautiful Chamois originates in the Canton of Berne. A nimble, surefooted animal, it is at home in the steep, rugged terrain of the European mountains, and can reach speeds of 10mph (17km/h) in these rocky landscapes. An ideal dairy goat, it produces good quantities of sweet-tasting milk, and its sweet and docile temperament makes it an ideal pet (until you try to trim their hooves - then they can be quite a handful!). They are, so I am told, brilliant bramble-mowers, and (something close to my heart) the milk of a herd of Chamois goes to create the beautiful and creamy Fryberg-Chäs cheese.

Dutch Landrace

Native to:
Netherlands

Now found:
Netherlands

Description

Preferably long-haired, the most common colours for this breed are brown black, blue grey and white. The horns are backwards with an outward twist. It weighs 58-90kg.

The Dutch Landrace is similar in many ways to the other breeds of northwest Europe. The breed has long been used to develop and improve many other breeds, and over the years the bloodline became weaker. By 1958, only two remained and, as usual, it was left to the enthusiasts to come to the rescue.

The milk yields of the Dutch Landrace are not sufficiently high for the breed to be farmed commercially, but the high-quality milk is used for cheesemaking. Herds of Dutch Landrace are used on National Reserve land to keep the moors and open grass areas free of trees.

Fainting Goat

Native to:
Unclear

Now found:
Marshall County, Tennesee

Description

Usually black and white or red and white, but occurring in the whole range of colours, this breed's coat which varies from short to shaggy. The horns sweep upwards and outwards, and the ears are horizontal to the head. Its height is 43.18-63.5cm (17-25"); its weight is 36.28-68kg (80-150lbs).

In the early 19th century an old farm labourer wearing strange clothes arrived in Marshall County, Tennessee accompanied by four goats and a cow - enter John Tinsley. Soon, everyone was talking about John's goats. If surprised or frightened, they stiffened or fell over; a condition known as mytonia congenita in which muscles briefly contract. This does not harm them in any; they will continue to chew the food in their mouths. Before moving on, John sold his goats to Dr H. H. Mayberry, who bred them and sold them locally. They soon became known as the Tennessee Fainting Goats. The breed takes up to three years to mature and makes an ideal pet.

Girgentana

Native to:
Aghanistan and Balochistan

Now found:
Sicily

Description

This is all-white, occasionally with brown spots on the face. It has upright corkscrew horns, 50cm or longer. Its height is 80-85cm (31.5-33.5); its weight is 50-65kg (110-143lbs).

An ancient breed with ancestors believed to have originated in Afghanistan and Balochistan, it arrived in Mazaro, Sicily when the Arabs invaded in 827. Despite further invasions, the breed prospered and slowly spread throughout the island. It is mainly found in the Province of Arigento, the old Gergenti from where the breed got its name. The Girgentana is now at serious risk of becoming extinct; in the mid-20th century there were in excess of 30,000 animals – now, the figure is nearer 500. Work is underway to ensure the survival of the breed. Now for matters close to my heart - food. The milk, apart from being good to drink, produces quality cheeses - either rolled in herbs or aged in Nero d'Avolo wine. Traditionally, the milk was used for children and the elderly.

Golden Guernsey

Native to:
Guernsey

Now found:
Guernsey; few further afield

Description

The long-haired coat of this breed comes in all shades of gold, from pale to bronze, sometimes with small white markings and a star on the forehead. It weighs 54.5-68kg (120-150lbs); its height is 66-71cm (26-28").

There have always been goats on Guernsey, the Golden Guernsey being indigenous to the island. Even during the dark days of World War II, when for five years the island was under German occupation, the breed was still being registered. One of the largest and possibly most well-known herds was that of Miss Milbourne of L'Ancresse, who owned over 50. This herd played an important part in helping to revive the breed in the 1930s. The gene pool is very small, and tremendous efforts are being made to improve the situation and secure the future of this beautiful breed. In an effort to safeguard their future, a number are being raised in New York State; but only the purebred with UK registration are allowed the proud title of Golden Guernsey.

Icelandic

Native to:
Iceland

Now found:
Iceland

Description

This has a long, coarse outer coat with guard hairs, and a beautifully soft cashmere undercoat. It is brown, grey, black and white, in various patterns. It can be horned or polled. It weighs 35-80kg (77-176lbs).

There are no pure Icelandic goats outside of Iceland. The Icelandic goat, also called the Settlement goat, arrived in Iceland with the Norwegian settlers over 1100 years ago. There have been no further imports since then. In 1986, six were exported to Scotland for a cashmere breeding programme. They became a quarter of a new synthetic goat breed, 'The Scottish Cashmere Goat'. In the early 20th century, there were in the region of 3000 Icelandic goats, but by the latter end of the century the number had plummeted to less than 400. The Icelandic government has introduced a conservation programme, and numbers are slowly increasing. At present, they are kept as pets, but their milk, cashmere and even ice cream have commercial and economic potential.

18.

Kiko

Native to:
New Zealand (originally Europe)

Now found:
New Zealand, USA

Description

The summer coat of the Kiko is smooth and shiny. It is usually white, although it can be coloured. In winter, the hair is long and flowing. The spiraling horns sweep upwards and outwards.   There is no published data on its typical height and weight.

The Europeans who discovered New Zealand in 1769 brought goats with them; predictably, a number of these escaped and thrived. These goats were to become the modern Kiko, which grew in number, having no natural predators. With no shelter, no supplementary feed, no veterinary care and no help giving birth, they became a tough and hardy breed, resistant to disease, parasites and the weather. The Kiko (Maori for 'meat') became totally self-sufficient. The breed, as it is known today, was developed by Garrick and Anne Batten on the South Island in the 1970s; only the best of the best were selected. In the early 1990s a number of Kiko were imported into the USA by Dr. An Peischel, where they continue to grow in popularity.

Kinder

Native to:
United States

Now found:
United States

Description

With short, fine hair in the full range of goat colours and patterns, the Kinder goat has longish ears that stick out to the sides and are horned. Its height is 66-71cm (26"-28"); its weight is 50-68 kg (110-150lbs).

The Kinder (KIN-der) is a cross between Nubian and Pygmy goats. In 1985 there was a death on the Zederkamm farm; the Nubian buck died. The owners, Pat and Art Showalter, were left with two Nubian females and no mate, but did not want to send the goats to another farm to mate. Art and Pat also bred Pygmy goats. The Pygmy buck bred with the Nubians; there were logs and rocks littering the field, so I will leave this to your imagination. In late June 1986, three Kinder does were born, and a year later, the first Kinder buck. The Kinder are gentle, family-friendly, and make good pets. The milk is ideal for drinking and for making yoghurt and cheese.

Kri-Kri

Native to:
Crete

Now found:
Crete

Description

This has a short, light brownish coat in summer, and a thicker coat in winter. It has a black line down the spine from the shoulder to the base of the tail and around the neck, and a white band above the knees. It has large horns. There is no published data regarding its typical height and weight.

The Kri-Kri is a wild goat found only on the isle of Crete and three small offshore islands as protection against extinction. Crete is not thought to be its homeland; it is believed to have been imported by the Minoans over 10,000 years ago. It is the largest wild animal on the island, with large, ringed horns of up to 90cm in length - with a ring for each year of the goat's life. By the 1960s, less than 200 Kri-Kri remained, grazing the highest mountains, up to 2,400m (8,000ft), of what is now a National Park. Through a preservation programme the numbers have grown to 2,000. They are still hunted despite hunting being strictly prohibited.

LaMancha

Native to:
United States

Now found:
United States

Description

Think of any goat colour or combination of colours, and that is the LaMancha. The hair is short, fine and glossy, and the ears are small - less than 2.5cm (1"). The goat's height is 71-76.2cm (28-30"), and they weigh 59-70.3kg (130-155lbs).

LaMancha dairy goats are recognizable by their very short ears. Short-eared goats are mentioned in ancient Persian writings, and the Spanish missionaries who colonised California in the mid-18th to mid-19th century brought with them short-eared goats. It is believed that these were the ancestors of the LaMancha. The first LaMancha were bred in 1938 by Mrs Eula Fay Frey; two offspring, Peggy and Nesta, are the foundation of the breed. The LaMancha are known for their high production of butter-fat rich milk which is used for making cheese, yoghurt, ice cream and soap. It is also the only goat breed developed in America. They have a laid-back attitude to life, are hardy and make superb brush and bramble mowers.

Messinese

Native to:
Sicily

Now found:
Sicily

Description

This goat has long hair in white, brown, black or red (in varying shades), and can be pied or streaked in these colours. Both males and females have horns. Its height is 67-72cm, and it weighs 38-55kg.

The Messinese (or Nebrodi) is a local breed of 'Capra' found in the mountainous regions of Peloritani and Nebrodis in the province of Messina and Etna in Sicily. The Messinese is well-adapted to living in these hilly areas, where the tendency is to let them roam free in an almost semi-feral state. The Messinese is bred especially for its milk, in an area that has an ancient tradition of cheese production by artisan cheese makers, whose skills and techniques are passed down from father to son. The Messinese feeds on the sparse mountain vegetation, browsing on basil, fennel, sorrel and rosemary – these herbs flavour the excellent range of cheeses produced.

23

Nigora

Native to:
United States

Now found:
United States

Description

This breed can be any of the range of goat colours and markings. It has erect ears and can be horned or polled. Its height is 48.26-73.66cm (19-29").

The Nigora is a small- to medium-sized goat developed in the USA in the1990s to produce both quality fibre and milk. The first doe (named Cocoa Puff of Skyview - good grief!) was black with black-and-white trimmings. Since then, there have been gentle improvements to the breed. It was developed with the urban hobby farmer and smallholder in mind, and its small size and relative ease of care also makes it an ideal pet. The fibre is classed mainly as 'cashgora' - a cross between cashmere and angora. The fibre is in three types, (A) having more mohair characteristics, (B) being the cashgora type, and (C) being the cashmere type. Quality milk and the makings of a cashmere sweater - what more could you ask for?

24

Nubian
(Anglo-Nubian)

Native to:
Middle East; developed in Britain

Now found:
Worldwide

Description

The short, fine, glossy coat can be any colour or pattern. This goat has a Roman nose and long, floppy ears. It weighs 61-79k (135-175lb), and its height is 76-89cm (30-35").

The Nubian was developed by humans, and originates in the Middle East; that they ever lived in Nubia is debatable. Arriving in the British Isles from France in 1883, it was crossed with the Old English Milch Goat. Improvements have been made over the years, with a little of this and a little of that, including a touch of Swiss. The Anglo-Nubian was officially recognized as a breed in 1896. The Nubian is also known for producing high quality, high butterfat milk - the 'Jersey' of the dairy goat world. Leaves and branches, including your herbaceous borders, are their foods of choice; eating grass is way down the list. If you are considering keeping a 'Lop-eared goat' a strong enclosure is required - Houdini is their middle name.

25

Peacock

Native to:
Swiss, German and Austrian Alps

Now found:
Swiss, German and Austrian Alps, but rare

Description

This has a thick coat, mainly white, with a black rear end and boots. The face has black stripes and spots. It has large horns. Its height is 73-80cm (28.5-31.5"), and it weighs 56-75kg (123.4-165.3lbs).

The Peacock goat is an ancient mountain breed; it was 'discovered' in 1887 but had been around in the Swiss, German and Austrian Alps for many years. The name came about because of a journalist's slip of the pen; he should have written Pfavenziege (striped goat) - but I think Peacock is more suitable for this beautiful animal. Blood tests carried out in the 1930s put the Peacock as a descendant of the Grisons Striped goat; more recent blood tests do not agree. The Peacock is docile, agile and hardy; well able to survive alpine weather and meagre pastures. It is on the Domestic Animals Breed List, and many associations are actively promoting it to ensure it does not disappear into the alpine mists.

26

Poitou

Native to:
France

Now found:
France

Description

With colours ranging from light to dark brown to blackish, the Poitou has long hair on the back and thighs and a white underbody. The head is dark, with white lines each side of the muzzle tip up to the ears. It can have horns, but does not always. It weighs 55-72kg (121-158.7lbs); its height is 75-80cm (29.5-31.5").

The legend of the Poitou dairy goat goes back to the year 732, when Charles Martel (Statesman and military leader) and the Duke of Aquitaine defeated the Saracens at Poiters. The goats left at the battle site are the forefathers of the Poitou goat. The goat we know today originated around the area of the Sevre River in the 1800s. In 1920, foot-and-mouth disease reduced the numbers from over 55,000 to almost zero. It took 50 years for the breed to begin recovery, but there are now almost 3,000. One use of the milk is to produce the famous Chabichou du Poitou cheese.

27

Pygmy

Native to:
West Africa

Now found:
Worldwide

Description

The full medium to long-haired coat comes in many shades of caramel, plus light, medium or dark grey. Any body colour is acceptable, and body patterns are created by the mixing of the many colours of hair. The billy has a long beard and a cape-like mane. Its weight is 24-39k (53-86lbs), and its height is 41-58cm at the shoulder.

The Cameroon Dwarf Goat, as it was originally known, comes from West Africa. During the 1950s many were imported into mainland Europe where they were exhibited in zoos as exotic animals. Within a few years they had found their way to the British Isles where they have proved very popular. The Pygmy is hardy and adapts to most climates. A great browser, it will clear all your weeds (and no doubt your herbaceous borders). An ideal pet for kids of all ages, they are gentle and affectionate and are frequently used as therapy animals.

28

Pygora

Native to:
United States

Now found:
Widespread

Description

With all Pygmy goat colours and markings including white, the colours of this breed vary through the seasons. The ears are erect, and it can be horned or polled. It weighs 29.48-43kg (65-95lbs). Its height is 48.26-73.66cm (19-29").

The inspiration for the Pygora came from Katherine Jorgensen seeing the coloured curly goats on a visit to the Navajo Indian Reservation. The first generation, a cross of the Pygmy and Angora, were bred in the mid-to-late 20th century, and were all white; the colours did not appear until the second and third generations. Many artists use this exquisite fibre for hand and machine spinning, knitting, weaving and tapestries. There are three types: (A) is an angora style - lustrous, curly and up to 6" long; (B) is curly, can be lustrous or matte, and is 3-6" long; (C) is a cashmere style, almost straight with a matte finish, and 1-3" long. Most Pygora, because of their docile and friendly nature, are kept as pets.

29

Rove

Native to:
France (heritage unclear)

Now found:
France

Description

The smooth, short, thick coat of this breed is mainly red or black, occasionally ash grey or red speckled with white. It has long, twisting horns. It weighs 60-90kg (110-220lbs).

The Rove is believed to have arrived in Marseille in 600BC, courtesy of the Phoenicians, after one of their ships foundered and a number of goats swam ashore. They were developed by local farmers, and were eventually named Rove after the village on the outskirts of Marseille. The most striking feature is the long, twisting horns, which can grow to 1.2m (4.0ft) in a mature adult. The Rove wanders the alpine countryside, eating aromatic herbs: citronella, thyme, and rosemary. These delicate flavours are to be found in the local goat cheeses. In the 1970s the breed was on the verge of extinction, but is now protected by the Association de Defense des Caprine du Rove, and is well on the road to recovery, with numbers now exceeding 6,000.

30

Saanen

Native to:
Switzerland

Now found:
Across Europe

Description

The hair of this breed is white or light cream. Some have a fringe of hair down the spine or over the thighs. The ears are erect and pointing forward. Bucks and does usually have horns. They weigh 68-91kg (150-200lbs), and their height is 76.2-101.6cm (30-40").

The breed originates in the Saanen valley in Switzerland. In the late 19th century, many thousands were rounded up and distributed throughout Europe, arriving in the British Isles via Holland in 1922. This is the largest of the dairy goats, and can produce up to 3.8l (1gal) of milk each day, gaining it the title of 'The Queen of Dairy Goats'. Like all goats, it needs a reasonable space to browse leaves and clover. Like me, they do not like getting wet, so a shelter from the rain and a good, warm, draught-free shelter for the winter are both musts. This calm, gentle, easy-to-handle animal makes an ideal pet and companion for children.

31

San Clemente

Native to:
Santa Catalina / San Clemente

Now found:
Worldwide, but rare

Description

With a light brown to dark red or amber coat, this breed has a black head with a brown stripe from around the eye to the muzzle, and a stripe running down the back. The ears stand out horizontally from the head. The horns flare up and outwards. Its height is 66-68.58cm (26-27"); it weighs 22.67-40.80kg (50-90lbs).

The goats were taken to San Clemente Island from Santa Catalina Island by Salvador Ramirez in 1875. In 1980 there were in excess of 15,000 goats on the island. The US navy is responsible for the upkeep of the island, and decided that the goats were damaging endangered plant species and island ecology. Helicopters were sent in to exterminate the goats, and thousands were slaughtered. The American Livestock Breeds Association intervened and a small number of breeding livestock were saved. These are shy, gentle creatures that make ideal pets and produce sweet-tasting milk. There are now only 250 worldwide. Don't we ever learn?

32

Savanna

Native to:
South Africa

Now found:
Canada

Description

This breed has short, smooth, white hair, occasionally with red, blue or black 'freckles'. The skin is loose and black. It has floppy ears and a slightly bent nose. The horns, preferably black, curve backwards and downwards. It weighs 56.7-90.7kg (125-200lbs).

The Savanna was developed in in 1955 in Douglas, South Africa by the Cillier family, using local goats. To survive in the harsh, local conditions (with extremes of heat and cold, and intense sunshine and rain) the goats needed to be tough, strong, disease- and parasite-resistant. In the cold, fine hair grows to give them added protection. Their diet is not that of the usual farm animal; they eat large bushes, trees and seed pods. To survive on this diet they have strong jaws, long-lasting teeth and strong back legs to enable them to reach the higher leaves and branches. If you want a breed of goat that doesn't require a wet-nurse, the Savanna is for you.

33

Spanish

Native to:
Spain

Now found:
United States

This breed has short hair, with longer hair on the lower parts of the body. Any colour is acceptable. The ears are long and fall to the sides of the head. The horns of the buck flare upwards and outwards.

In the 16th century, Spanish explorers brought goats from their homeland to the Caribbean islands. Eventually, they arrived in what is now the USA and Mexico. The Spanish goat of today is the result of cross-breeding with many of the New World breeds. Until the Mayflower anchored in 1662, the only goats in North America were the Spanish. It is an extremely hardy breed, but is unfortunately threatened with extinction, and is on the American Livestock Breeds Conservancy watch list. Like most goats, they were originally bred for food, but their other names ('brush' or 'scrub' goat) give a clue – they do an excellent job of clearing brush and unwanted plants from pasture land.

34

Stiefelgeiss

Native to:
Switzerland

Now found:
Switzerland

Description

This breed's colours range from light greyish brown to dark reddish brown. They have beards, and longer hair on their back legs - usually a different colour. They are horned. Their height is 67-85cm (26.5-33.5"); they weigh 50-80kg (110-176lbs).

The Stiefelgeiss (Booted goat) is a robust and hardy breed, well-suited to life in its harsh mountain habitat. Until 1920 it could be found in the uplands of St Gallen in Glarus, Switzerland. By the 1980s, the breed was on the verge of extinction. It was then that ProSpecieRara, which supports the breeding and cultivation of traditional animals and crops, took control. The Booted Goat Breeders Club of Switzerland has now taken over management of the breed. Farmers across Switzerland are actively being encouraged to breed the Stiefelgeiss for its milk, meat and fibre. Its appetite for leaves, buds and bark also make it an ideal tool to preserve the quality of pasture land, and females make ideal surrogate mothers.

35

Tauernsheck

Native to:
Austria

Now found:
Austria

Description

This is a short-haired, brown, black and white goat, with a white line on the face and large white spots on the sides of the body. It has black and brown legs. Both bucks and does are horned. It weighs 50-80kg (110-176lbs); its height is 70-90cm (29.5- 35.5").

The Tauernsheck is a very rare breed, its ancestry beginning in the 1800s with the Austrian Landrace, the Pinzgauer, another very rare breed from Asian ancestors and an unknown passing friend that was responsible for the white spots. They are found mainly in the Rauris valley in Austria and in the area around the Grossglockner Mountain. The colours are bred to design so they can be easily seen on the mountainsides. In 1994 an Austrian Government survey found only 200 Tauernsheck existed; these numbers are now slowly improving. The Tauernsheck spend summers high in the mountains browsing on the lush grass and breathing the fresh air. They are healthy and long-lived, and are excellent mothers. Twin kids are common.

36.

Toggenburg

Native to:
Switzerland

Now found:
Worldwide

Description

The medium-length coat of this breed is coloured from light fawn to darkest chocolate. Ears are white, each with a dark spot in the middle. The face has two white stripes. The lower parts of the legs are white. It is usually without horns. It weighs 56.7kg-136kg (125-200lbs); its height is 0.76m-0.96m (30-38").

Developed in the Toggenburg region of St Gallen in Switzerland over 300 years ago, this goat was the main source of income for poor families. Village Toggs were pooled and grazed on the Alpine pastures as one herd; the cheese produced from the sweet-tasting milk was distributed among the owners. The Togg was imported into the British Isles in 1822, and four were exported to America the following year. This robust breed has been exported to many countries, but like most Alpine breeds, is happier in temperate climes. A Toggenburg holds the Guinness World Record for producing g 5182.5lt (1140 gals) of milk in 365 days.

Valais Blackneck

Native to:
Switzerland

Now found:
Switzerland

Description

This breed has long, shaggy, wavy hair. It is white from the shoulders back; the head, neck and front legs are black. The horns are long, wavy and arched. It weighs 50-70kg (110-154lbs); its height is 75-80 cm (29.5-31.5").

The Valais Blackneck is a dual-purpose breed with an international reputation. It is known by many names, including Col Noir de Valais and the Glacier goat. Developed from indigenous goats crossed with the Italian Kupferziege goat, it was further improved by selective breeding. It is found mainly in Lower Valais, Switzerland. This hardy breed has the ability to tolerate harsh winter conditions, but is stabled in the worst winter weather. A gourmet browser, it feeds only on the greenest alpine grass, fresh herbs, and flowers, and produces up to 2 litres of milk daily. In 1970, this protected species was close to extinction; fortunately, numbers of this most beautiful, most-photographed goat are improving. There are, at present, approximately 3000.

Goat Talk

A baby goat is called a kid.

A female goat is called a nanny or a doe.

A male goat is called a billy or a buck.

A kid can recognise its mother's call soon after birth.

Goats have excellent co-ordination.

Goats can balance on precarious mountain ledges.

Goats can climb trees.

Many goats can jump over 152cm (5').

'Goat' is one of the 12 Chinese zodiac signs.

Goats have excellent night vision.

Goats are believed to have discovered coffee in Ethiopia. They became more active after eating the fruit of a certain bush. The herder tried them, jumped up and down, and said 'I like this!'

Some feral goats in Scotland are the descendants of goats abandoned during the Highland Clearances.

Goats are good swimmers.

The milk of each breed of goat has a different taste.

Goats can catch cold and suffer from pneumonia.

There are over 200 breeds of goat.

There are over a billion goats worldwide.

More people eat goat meat and drink goats' milk than eat meat and milk from other animals.

Goats should not be kept alone; they thrive on company.

Many goats can live for up to 12 years, and some can live for up to 15 years.

Goats are often kept as companions to racehorses for their calming effect. If you wanted to nobble a horse you would remove the goat companion. The horse would become unsettled and run badly, hence 'getting his goat'.